GEISEL LIBRARY
UNIVERSITY OF CALIFORNIA, SAN DIEGO
LA JOLLA, CALIFORNIA

AF683853

UNIVERSITY OF CALIFORNIA, SAN DIEGO
LA JOLLA, CALIFORNIA
3 1822 02663 7534

DOCUMENTED BRIEFING

RAND

Improving Support to CINC Theater Engagement Plans

Phase I

Roger A. Brown, Leslie Lewis, John Y. Schrader

Prepared for the
Joint Staff

National Defense Research Institute

Approved for public release; distribution unlimited

The research described in this report was sponsored by the Joint Staff. The research was conducted in RAND's National Defense Research Institute, a federally funded research and development center supported by the Office of the Secretary of Defense (OSD), the Joint Staff, the unified commands, and the defense agencies under Contract DASW01-95-C-0059.

ISBN: 0-8330-2970-3

The RAND documented briefing series is a mechanism for timely, easy-to-read reporting of research that has been briefed to the client and possibly to other audiences. Although documented briefings have been formally reviewed, they are not expected to be comprehensive or definitive. In many cases, they represent interim work.

RAND is a nonprofit institution that helps improve policy and decisionmaking through research and analysis. RAND® is a registered trademark. RAND's publications do not necessarily reflect the opinions or policies of its research sponsors.

© Copyright 2001 RAND

All rights reserved. No part of this book may be reproduced in any form by any electronic or mechanical means (including photocopying, recording, or information storage and retrieval) without permission in writing from RAND.

Published 2001 by RAND
1700 Main Street, P.O. Box 2138, Santa Monica, CA 90407-2138
1200 South Hayes Street, Arlington, VA 22202-5050
RAND URL: http://www.rand.org/
To order RAND documents or to obtain additional information, contact Distribution Services: Telephone: (310) 451-7002; Fax: (310) 451-6915; Internet: order@rand.org

PREFACE

This documented briefing contains the findings and recommendations resulting from the Phase I study on improving support to Commander-in-Chief (CINC) Theater Engagement Plans (TEPs). The material presented in this document was used to brief the Global Engagement Working Group Conference that was held at RAND's Arlington, VA office in August 2000 and formed the briefing of the concept for improving TEP given by the J-5 to the CINCs in September 2000. The project is cosponsored by the Director for Strategic Plans and Policy (J-5) and the Director for Force Structure, Resources, and Assessment (J-8) in the Joint Staff.

The purpose of the Phase I study is to help the J-5 and J-8 develop a concept for improving the engagement mission's visibility and, ultimately, resource support. The objectives are to develop a framework for assessing global engagement and its associated resource demands. The framework will improve visibility of engagement resource shortfalls and the mission's relative resource priorities. Phase II will flesh out the concept, obtain additional insights on the complexity of the engagement mission, and develop better connectivity between the engagement mission and both the Joint Strategic Planning System (JSPS) and the Planning, Programming, and Budgeting System (PPBS).

This research will be of interest to the Office of the Secretary of Defense (OSD), the military departments, the CINCs and their staffs, the Joint Staff, and others concerned with national security policy and defense resourcing.

The Engagement Strategy Branch within the J-5's Strategy Division and the Forces Division within the J-8 provided project oversight. Within RAND, the work was conducted within the Strategy and Policy Division, part of RAND's National Defense Research Institute (NDRI), a federally funded research and development center sponsored by the OSD, the Joint Staff, the unified commands, and the defense agencies.

CONTENTS

ACKNOWLEDGMENTS

This documented briefing could not have been produced without the cooperation and involvement of many individuals. The authors are grateful to the directors and staffs of the J-5 and J-8. In particular, we thank our project sponsors, LTG Edward Anderson (USA), Director of the J-5, and Lt. Gen. Bruce Carlson (USAF), Director of the J-8, for involving us in the project. We also extend our appreciation for the support, guidance, and assistance we received from Rear Admiral Barry Costello (USN), Deputy Director for Strategy, J-5, and COL (P) James Coggin (USA), Executive to the Director, J-5. Our special thanks go to the members of the Engagement Branch, Strategy Division, J-5, Commander Robert Presler (USN), Lt. Col. Charles Lutes (USAF), and Lt. Col. Lynn Herndon (USAF), for their guidance, support, and assistance throughout Phase I and the development of Phase II requirements. We also extend our special thanks to Lt.Col. David Elwing (USMC) from Forces Division, J-8, who assisted us in obtaining critical information and provided useful insights into the overall project.

We also extend our appreciation to our research assistants, Kenneth Myers and Deanna Weber, who helped us compile, format, edit, and write the final report. Ms. Weber also set up and managed the August 2000 Engagement Conference at RAND.

Finally, we wish to acknowledge our colleague Bruce Pirnie, whose useful reviews led to improvements to the briefing.

The authors, of course, are completely responsible for any errors in the briefing or shortcomings in the research.

ACRONYMS

AA	Activity annex; Appropriation Apportionment
CINC	Commander-in-Chief
CJCS	Chairman, Joint Chiefs of Staff
CJCSI	Chairman, Joint Chiefs of Staff Instruction
CJCSM	Chairman, Joint Chiefs of Staff Manual
CPA	Chairman's Program Assessment
CPG	Contingency Planning Guidance
CPR	Chairman's Program Review
DCSAS	Deputy Chief of Staff for Air and Space Operations
DoD	Department of Defense
DPG	Defense Planning Guidance
DUSA-IA	Deputy Undersecretary of the Army for International Affairs
EUCOM	European Command
FYDP	Future Years Defense Plan
GNFPP	Global Naval Force Presence Policy
IAP	International Affairs Plan
IPL	Integrated Priority List
IPR	In-Process Review
ISA	International Security Affairs
JPD	Joint Planning Document
JSCP	Joint Strategic Capabilities Plan
JSPS	Joint Strategic Planning System

JSR	Joint Strategy Review
JV	Joint Vision
MOE	Measure of effectiveness
MTW	Major Theater War
NMS	National Military Strategy
NSS	National Security Strategy
OPLAN	Operational Plan
ONW	Operation Northern Watch
OPTEMPO	Operational tempo
OSD	Office of the Secretary of Defense
OSW	Operation Southern Watch
PA&E	Program Analysis and Evaluation
PB	Program Budget
PBD	Program Budget Decision
PDM	Program Decision Memorandum
POM	Program Objective Memorandum
PPBS	Planning, Programming, and Budgeting System
PRG	Program Review Group
PRO	Prioritized Regional Objective
PWG	Policy Working Group
QDR	Quadrennial Defense Review
RBA	Revolution in Business Affairs
RMA	Revolution in Military Affairs
SC	Strategic concept
SecDef	Secretary of Defense
SSC	Smaller-scale contingencies
TEP	Theater Engagement Plan
TEP AA	Theater Engagement Plan Activity

	Annex
TEP SC	Theater Engagement Plan Strategic Concept
TEPMIS	TEP Management Information System
TPFDD	Time-phased Force Deployment and Development
UFR	Unfunded Requirements
USC	United States Code
USCENTCOM	United States Central Command
USEUCOM	United States European Command
USFK	United States Forces Korea
USSOCOM	United States Special Operations Command
USSOUTHCOM	United States Southern Command
WWPC	World Wide Planner Conference

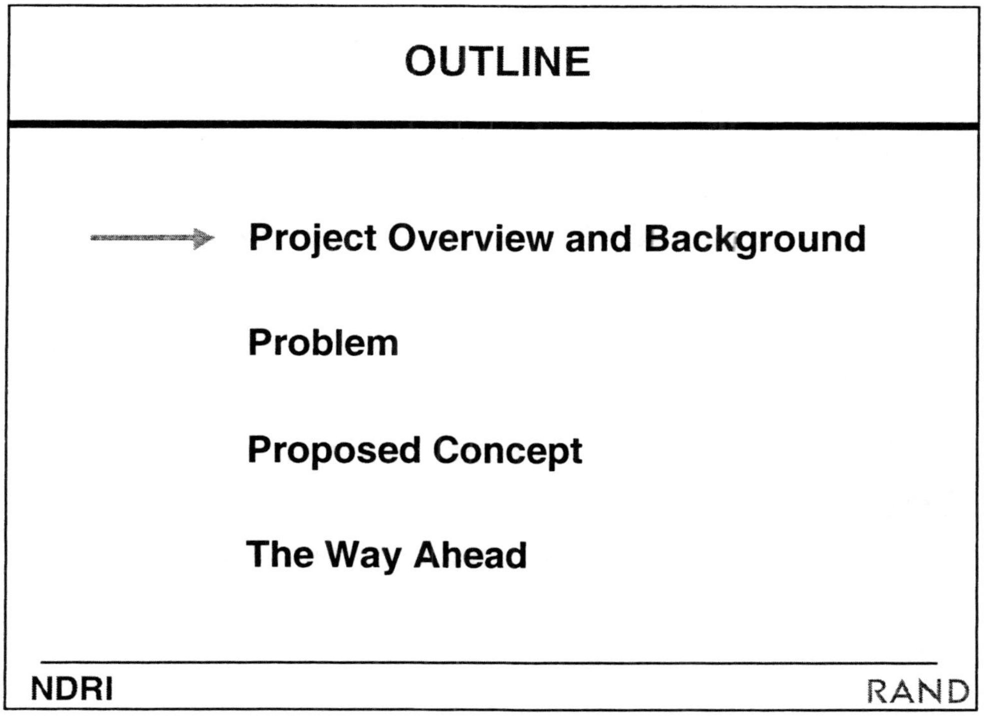

Figure 1.1—Outline

PROJECT OVERVIEW AND BACKGROUND

Outline

This documented briefing is organized into four major sections (see Figure 1.1). This section provides an overview of the study, including its background, objectives, and the organization of the project into two phases.

Background

The foundation for this study to assist the Joint Staff in improving support to Commander-in-Chief (CINC) Theater Engagement Plans (TEPs) derived from the February 2000 CINCs' Conference. At this conference, the CINCs tasked the Joint Staff with reviewing the TEP process for value added versus effort required as well as with evaluating that process to determine how to link resources and requirements. The Joint Staff asked RAND to assist them in the development of a concept to improve the visibility of the engagement mission and to examine ways to improve resource allocation to TEPs.

Two-phase project cosponsored by J-5/J-8

- **Phase I was initiated in mid-July to develop the concept and briefing for the September CINCs' Conference**
- **Phase II will flesh out the concept, obtain additional insights, develop process connectivity, provide recommendations for change, and recommend an implementation plan**

Objectives:

- **Evolve Theater Engagement Plan (TEP)**
  - **Develop a framework for assessment of global engagement demand (J-5)**
  - **Improve visibility of engagement resource shortfalls and relative priorities (J-8)**
- **Improve resource support for engagement**

NDRI RAND

Figure 1.2—Project Overview

Project Overview

The engagement project is cosponsored by the Director for Strategic Plans and Policy (J-5) and the Director for Force Structure, Resources, and Assessment (J-8). Phase I was initiated in mid-July 2000 to help the Joint Staff develop a concept for how the engagement mission might be addressed in a more integrated way. The work for Phase I was compressed into a five-week period so that project sponsors could present an initial concept in early September 2000 at the CINCs' Conference (see Figure 1.2).

Following agreement among the Joint Staff, services, and CINCs on the RAND conceptual approach laid out in Phase I, Phase II will further flesh out the concept, obtain additional insights from all the players involved, develop recommendations for improvement, and assist in the preparation of an implementation plan. Both Phase I and Phase II of the work necessitate extensive interaction and collaboration with all the participants involved in theater engagement planning—the Office of the Secretary of Defense (OSD), CINCs, Joint Staff, and services.

The J-5 and J-8 sponsors were very clear at the initiation of Phase I of this project that two objectives had to be attained. The first was that the TEP needed to be improved so that it provided a mechanism for the J-5 to gain a more integrated and structured way to understand the total demands of the engagement mission. To attain this objective, some type of framework for assessment had to be developed. As part of this objective, the proposed approach was also expected to improve the visibility of the total engagement mission as well as the resource issues associated with that mission. In particular, the J-8 sought a means by which engagement resource shortfalls and their relative priority against the totality of CINC resource demands could be evaluated.

The second objective was that at the conclusion of the project, the Joint Staff had to improve the potential for engagement resource support to the CINCs. The second objective would build on the efforts employed to achieve the first objective by enhancing the connection of the theater engagement mission to the principal DoD resource decisionmaking process.

- **How does engagement fit within the spectrum of CINC missions?**
- **What guidance do the CINCs receive to conduct the engagement mission?**
- **What requirements are received from the CINCs to resource the engagement mission?**
- **How do the DoD's resourcing processes support engagement (e.g., JSPS, PPBS)?**
- **What are the MOEs and feedback mechanisms for assessing engagement resources and execution success?**
- **What is the basis for allocation of forces?**
- **How might global engagement be assessed?**
- **How might the process be improved, building on existing processes and capabilities (e.g., JSPS, PPBS, TEPs, TEPMIS)?**

NDRI RAND

Figure 1.3—Phase I Assessment Issues

Phase I: Assessment Issues

The cosponsors requested that RAND sponsor an expanded Policy Working Group (PWG) in mid-August 2000 both to gain insights into the issues associated with the theater engagement mission and to share some initial insights. The PWG consisted of representatives from CINC planning staffs and representatives from the OSD, Joint Staff, and services involved in planning and programming.

To provide some initial structure to the Phase I work, RAND developed several assessment issues. These issues were also designed to facilitate the PWG discussions.

Building on prior RAND work with United States Special Operations Command (USSOCOM) and with United States Forces Korea (USFK),[1]

[1] See John Y. Schrader, Leslie Lewis, William Schwabe, C. Robert Roll, and Ralph Suarez, *USFK Strategy-to-Task Resource Management*, Santa Monica: RAND, MR-654-USFK, 1996;

RAND developed a set of issues that defined the problems, the players' responsibilities, and the operational context of the engagement mission. We also sought different perspectives on how the engagement mission fit within the broader spectrum of CINC missions, Joint Staff activities, and service Title 10 United States Code (USC) functions. In addition, we sought to gain insights on how the engagement activity was addressed in formal DoD decisionmaking processes. These processes are the Joint Strategic Planning System (JSPS) and the Planning, Programming, and Budgeting System (PPBS).[2]

Critical to the development of a framework and concept for evaluating global engagement requirements and resources is the use of existing engagement-planning and data-gathering processes and databases. Regional CINCs develop TEPs, which are designed to provide the CINCs' with a strategy for supporting their respective Prioritized Regional Objectives (PROs) and the planned sets of specific activities associated with the engagement effort in that theater for each fiscal year. Figure 1.3 provides a sampling of issues indicating the breadth and scope of issues associated with the engagement mission and its related resources.

---

and Leslie Lewis, James A. Coggin, and C. Robert Roll, *The United States Special Operations Command Resource Management Process*, Santa Monica: RAND, MR-445-A/SOCOM, 1994.

[2]The JSPS is a process managed by the Chairman, Joint Chiefs of Staff (CJCS). Several subprocesses are associated with the JSPS; they are carried out over a two-year period. The JSPS provides strategic assessments, strategic direction, and strategic plans and advises the Secretary of Defense (SecDef) on programming issues associated with operational readiness. See Chairman, Joint Chiefs of Staff, CJCS Instruction, *Joint Strategic Planning System*, CJCSI 3100.01A, September 1999.

The PPBS is a cyclic process containing three distinct but interrelated phases. The JSPS informs the planning and programming phases of the PPBS. The PPBS is the dominant resource management and decisionmaking process within the DoD. The result of the PPBS process is the development of service programs that identify how the services will meet the guidance laid down by the OSD in the Defense Planning Guidance (DPG). Through a series of forums, the service programs are adjudicated and direction is given on how the service programs will be modified to meet the OSD's guidance. After a series of budgetary reviews, the SecDef submits the DoD's inputs to Congress for final review, approval, and allocation of funds. The final decisions are contained in the President's budget. For information on all key DoD decisionmaking processes, see Naval War College, *Resource Allocation, Volume 1, The Formal Process*, 1996.

- Applied viewpoint vs. vantage point approach to:
  - Identify roles and activities of participants in engagement missions (e.g., internal and external to DoD)
  - Determine complexity of mission
- Employed economic model of supply, demand, and integration to determine and assess interrelationships in planning and resourcing engagement
  - Analyzed JSPS and PPBS for connectivity
  - Analyzed roles of key players in engagement mission
- Used RAND's strategy-to-tasks-to-resources as an analytic template to define an approach
- Used TEPs and TEPMIS as the foundation for defining the scope of the engagement mission

NDRI RAND

Figure 1.4—Analytic Approach

Analytic Approach

The research approach RAND used was to quickly gain sufficient knowledge of some of the varying views on the engagement mission to frame the key elements of the problem. Toward this goal, the technique known as "viewpoint–vantage point" was used as described in Figure 1.4. This approach provides a structured means by which one is able to identify the roles and various perspectives of the players involved in the engagement mission. The top-down and bottom-up aspects are parts of the approach that enable insights to be gained on how the different players view the mission and problems, their perspectives of their responsibilities, and how these perspectives might vary from one player to another.

The economic model of supply, demand, and integration was also used to determine and assess how the interrelationships operate. Particular attention was given to the interrelationships and interactions between the planning of the engagement mission (demand) and how that mission was resourced (supply). Also addressed was the issue of who is performing the integration of the mission in terms of balancing supply against demand. The economic model was also used to identify the hierarchical relationships among the various players. For example, the OSD, services, and Joint Staff operate at the highest DoD level of the model. The OSD sets the demand, the services operate as the suppliers of resources, and the Joint Staff (J-5/J-8) along with the OSD—primarily OSD Program Analysis and Evaluation (PA&E)—perform the integration function. At the subordinate level, however, the CINCs, service components, Joint Staff, and service headquarters also perform different functions associated with demand, supply, and integration. These complex relationships needed to be identified and understood.

The RAND study team developed a strategy-to-tasks-to-resources analytic framework to determine a level of focus that would be most useful for each player in the two key DoD decisionmaking processes (JSPS/PPBS).[3] This framework also provided a mechanism by which the project team could assess how each player participates in the JSPS and PPBS processes. Finally, because the framework focuses on the linkage of warfighting objectives to capabilities, it provided a structure that facilitated the assessment of how engagement operates or might operate within the larger operational context of CINC missions and the engagement mission's impact on resources.

RAND used the existing TEPs and the evolving TEP Management Information System (TEPMIS) as key elements to define the foundation of the engagement mission.

---

[3]The strategy-to-tasks framework was developed at RAND during the late 1980s and is currently being used by several DoD organizations. The framework is designed to link resource decisions to specific military tasks that require resources, which in turn are linked downward hierarchically from higher-level operational and national security strategies to supporting programs and tasks. See Leslie Lewis and C. Robert Roll, *Strategy-to-Tasks: A Methodology for Resource Allocation and Management*, Santa Monica: RAND, P-7839, 1993.

- **Collected, reviewed, and assessed documents and data:**
  - **DPG, CPG, CJCSI, CJCSM, and JSCP**
  - **PROs, TEPs, and TEPMIS**
  - **Service POMs, relevant PRG issues**
- **Worked closely with staff from J-5 and J-8 to obtain perspectives and understand breadth of issues**
- **Reviewed ongoing engagement plans and activities**
- **Reviewed evolution of TEPMIS with EUCOM staff**
- **Conducted selected interviews (e.g., OSD ISA, services)**
- **Participated in Joint Staff TEP assessment tool workshop**
- **Conducted PWG engagement concept vetting conference (August 17)**
- **Prepared draft briefings for Director, J-5**

NDRI RAND

Figure 1.5—Completed Study Activities

What We Have Done

In Phase I, the project team turned first to published literature and guidance on engagement to gain some initial insights (see Figure 1.5). In Phase II, many of these initial sources of information will be reviewed in greater depth. We reviewed the Defense Planning Guidance (DPG), the Contingency Planning Guidance (CPG), the Chairman, Joint Chiefs of Staff Instruction (CJCSI) on the JSPS (CJCSI 3100.01A), the CJCS Manual (CJCSM) for Theater Engagement Planning (CJCSM 3113.01A), and the Joint Strategic Capabilities Plan (JSCP), all of which contain information on how the theater engagement mission is addressed within the JSPS and PPBS processes. Assessments were also done on the evolution, utilization, and relationships of PROs, TEP Strategic Concepts and Activity Annexes (TEP SCs and AAs), and TEPMIS.

The team also interviewed selected individuals on the Joint Staff who are familiar both with engagement and with associated planning and programming. For example, the action officers in the J-5 and J-8 who deal with the engagement mission were interviewed. The PWG Engagement

Conference provided the RAND team with another opportunity to gain additional insights from CINCs, the OSD, and service representatives. More extensive interviews will be conducted in Phase II. Additional insights were also gained from initial interviews with selected members of the services and OSD staffs—in particular, some members of the Army international affairs organization in the office of the Deputy Undersecretary of the Army for International Affairs (DUSA-IA) and Air Force regional planners in the office of the Deputy Chief of Staff for Air and Space Operations (DCSAS) on the Air Staff.

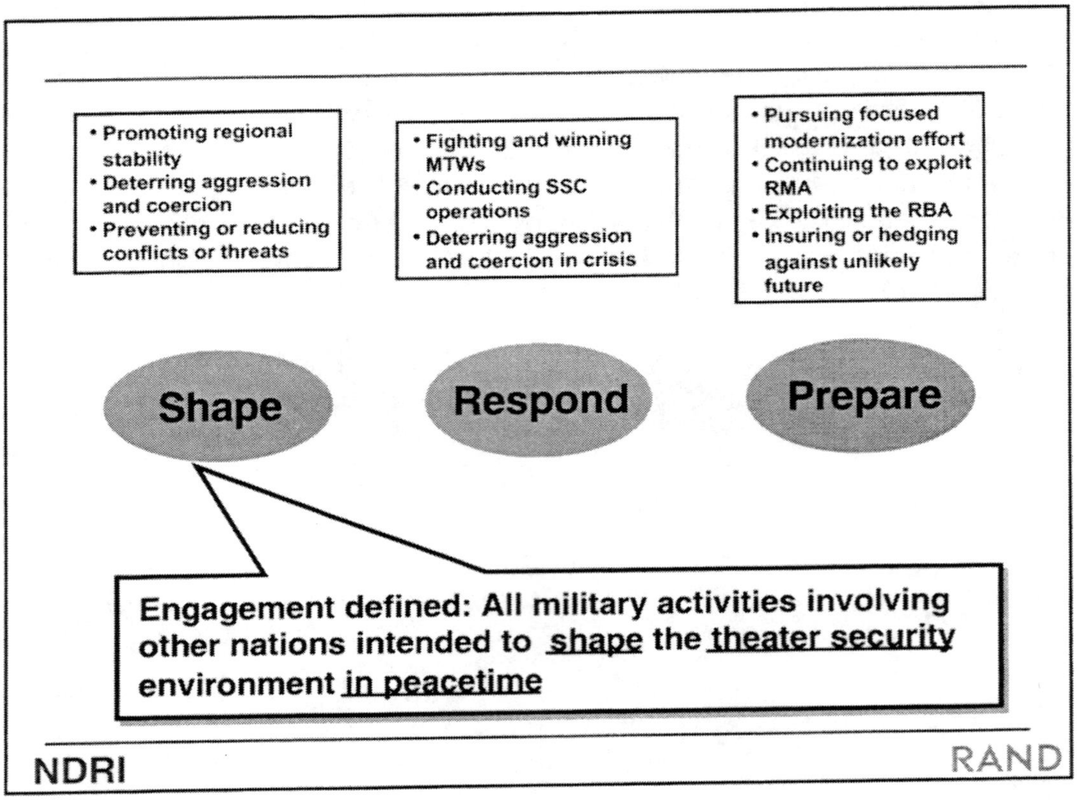

Figure 1.6—The Defense Strategy

The Defense Strategy

The defense strategy defines where engagement fits within the spectrum of DoD missions (Figure 1.6). It notes that engagement directly supports the shape pillar of the military strategy. The shape function encompasses the ability of the U.S. military to promote regional stability, deter aggression and coercion, and prevent or reduce conflicts.[4] Engagement is further defined as including all the military activities involving other

---

[4]The Defense Strategy derives from the 1997 Quadrennial Defense Review, which addresses the world today and tomorrow, identifies required military capabilities, and defines the programs and policies needed to support them. The Defense Strategy builds on the President's National Security Strategy (NSS) and determined that for the near and long term it must continue to shape the strategic environment to advance U.S. interests, maintain the capability to respond to the full spectrum of threats, and prepare now for the threats and dangers of tomorrow and beyond. See William S. Cohen, Secretary of Defense, *Report of the Quadrennial Defense Review*, 1997, and *Annual Report to the President and Congress*, 1999, pp. 5–6; and the *U.S. Department of Defense, Contingency Planning Guidance (CPG), Fiscal Years 1999–2003* (U), 1998.

nations that are intended to shape the theater security environment in peacetime.[5]

As will be seen in a later discussion, we believe that this definition is too narrow on the one hand and too broad on the other to facilitate a clear understanding of the totality of the engagement mission, how it is planned for, and how it is ultimately resourced. A common basis for planning and resourcing is needed that recognizes the varied demands of each CINC's regional focus and allows a single basis for assessment. The current definition does not provide a basis for the CJCS or the Joint Staff to perform the required assessment of engagement in a global context.

---

[5]Chairman, Joint Chiefs of Staff, Manual, *Theater Engagement Planning*, CJCSM 3113.01A, April 2000.

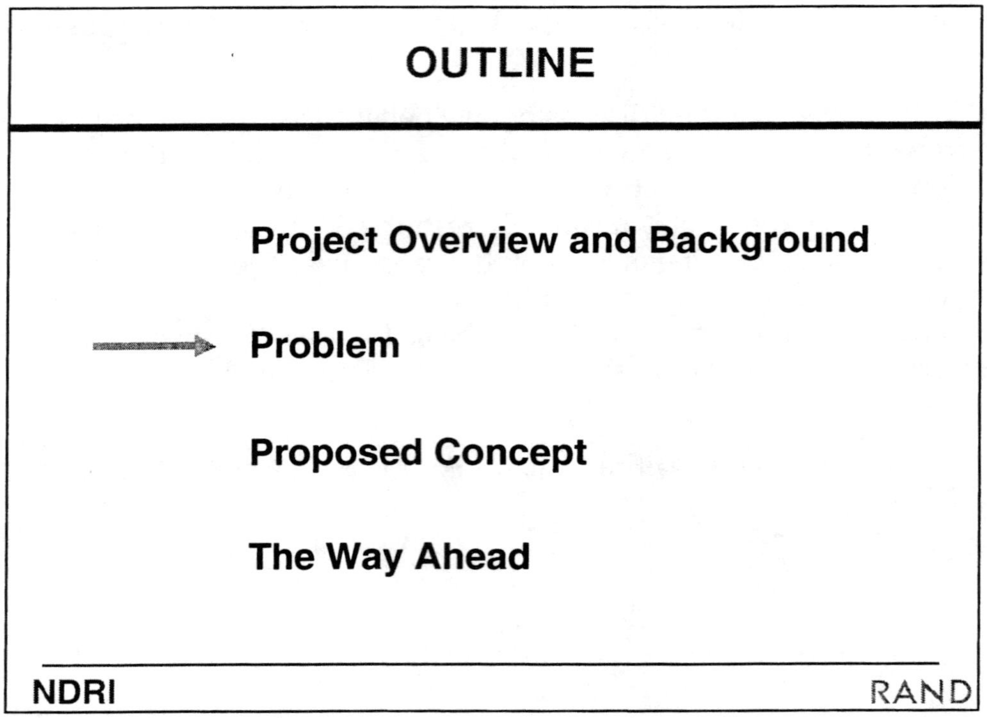

Figure 2.1—Outline

THE ENGAGEMENT PROBLEM

Outline

This section discusses the problems found during Phase I of the study in the planning and programming of the engagement mission (Figure 2.1).

Overview of the Engagement Problem

The engagement mission has been a CINC theater responsibility for many years. Attempts in the late 1990s to formally recognize the mission resulted in the definition of engagement-related activities in a narrow peacetime operational context, which has hindered its ability to compete in the formal DoD planning and programming processes. For example, the TEP process evolved within a separate and limited planning stovepipe lacking any connection to resources. The limits on resource support for TEPs in the JSCP are clear in that the engagement mission is not a basis for the assignment of forces.[6] Engagement also has incomplete links to the broader components of the JSPS and PPBS processes, thereby resulting in an inability to compete for resources in the major decision forums.

---

[6]Chairman, Joint Chiefs of Staff, Manual, *Instructional Joint Strategic Capabilities Plan (JSCP) FY 1998* (U), CJCSM 3110.01A, Appendix A to Enclosure E, Engagement, Planning Guidance, December 1998.

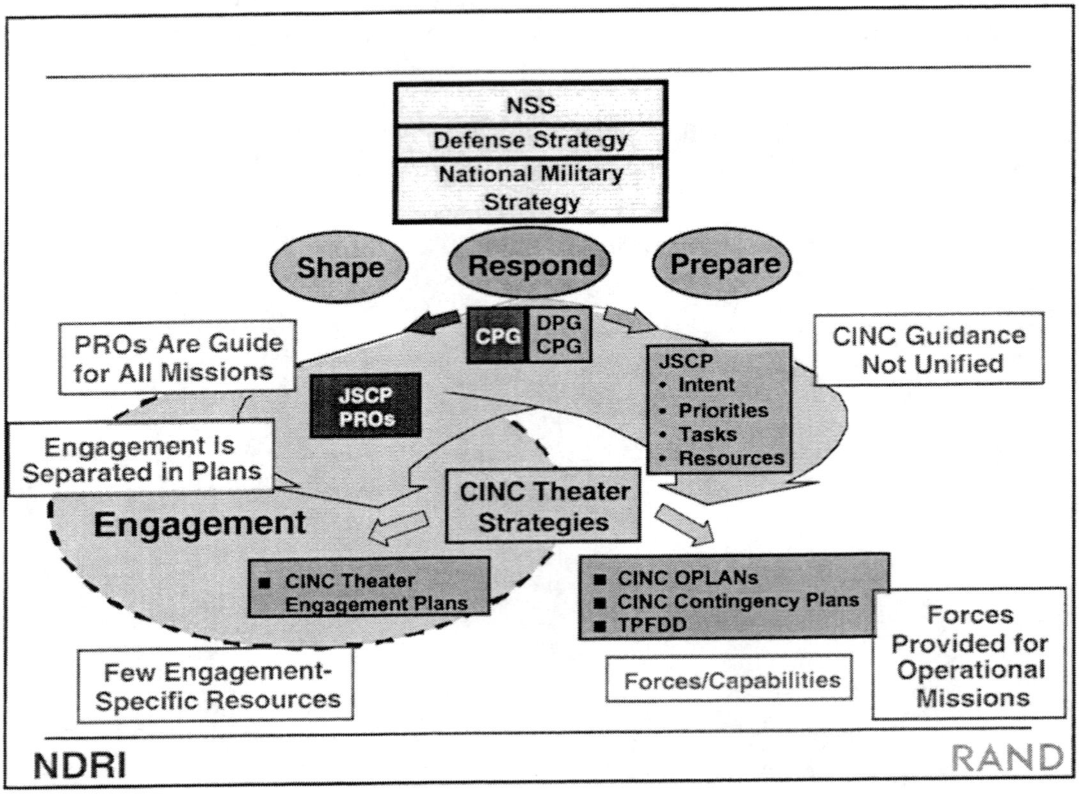

Figure 2.2—The Engagement Problem

Figure 2.2 depicts the current planning stovepipe in which theater engagement operates. It shows, for example, how the CINC theater strategic planning process is operating in a two-pronged, bifurcated system. The left-hand side of Figure 2.2 shows the guidance that comes from the OSD in the CPG and from the CJCS in the JSCP in the development of TEPs. The TEPs are focused on the shaping portion of the Defense Strategy without any strong connectivity to or provision for resources. In part, this problem emanates from the initial perception of the engagement mission as not demanding additional resources. In fact, TEP implementation was determined early on by the OSD to be highly dependent on the same capabilities and resources provided for the Respond and Prepare components of the strategy as they applied to other CINC missions.

The right-hand portion of Figure 2.2 shows the formal decisionmaking processes operating within the JSPS and PPBS. These integrated processes guide, direct, and resource the traditional CINC missions, resulting in the identification, assignment, and funding of forces and capabilities. Although it is recognized that the engagement mission is an enabler for other CINC missions, that mission is not directly connected to or funded within the traditional planning and programming pipelines.[7]

---

[7]Chairman, Joint Chiefs of Staff, Instruction, *Chairman of the Joint Chiefs of Staff, Commander in Chiefs of the Combatant Commands, and Joint Staff Participation in the Planning, Programming, and Budgeting System*, CJCSI 8501.01, Enclosures A and B, April 1999.

Four interrelated problems were identified; the fourth problem may not be solved by this study

(1) The engagement mission consists of everyday activities that facilitate National Security Strategy, but:

- It is viewed as an all-encompassing mission that cannot be planned for like traditional CINC missions
- It provides no basis for allocation of forces in the traditional sense

(2) The engagement mission is not connected to formal DoD decisionmaking processes

- Operates primarily within a planning process stovepipe
- JSCP provides little additional guidance on TEPs
- DPG:
  - Planning guidance instructs services only to support activities
  - Does not provide specific engagement programming guidance
- PPBS process has no direct resource feedback on TEP support
- Few visible or direct links from CINC TEPs to service POMs
- IPLs seldom raise major resource issues for engagement

NDRI RAND

Figure 2.3—Phase I Initial Insights (1-2)

Phase I: Initial Insights

The research team identified four major problems associated with the manner in which the engagement mission is currently planned and resourced (Figures 2.3, 2.4, and 2.5). First, engagement is an all-encompassing daily set of activities that enables the CINCs to assure their capabilities for operational missions—but it is a mission without additional forces or resources. The TEP is currently separate from the deliberate planning process in the JSPS.

Second, the engagement mission is weakly connected to the formal DoD decisionmaking processes, and therefore it operates primarily in a separate planning stovepipe that is not linked to programming and gets little direct visibility within the decisionmaking processes of the OSD or the services.

(3) TEPs' structure inhibits the Joint Staff's ability to link the engagement mission's objectives with capabilities

- TEPs are structured around the strategy-to-tasks concept as developed in CJCS Manual 3113.01A, but
- PROs are regionally focused national security objectives that do not provide a basis for engagement resourcing or assessment
- Resources at the lowest level of detail do not inherently link to operational objectives and capabilities (e.g., two SEALs and a raft) as seen with traditional CINC missions—no basis for priorities
- Linkages between the top level and lower levels of the hierarchy are asserted but not logically or analytically connected—difficult to defend

(4) Other federal agency programs supporting TEPs are uncertain

- Other federal agencies do not program future resources, forcing incomplete CINC planning until execution
- Variety of programs and numerous proponents complicate coordination

NDRI RAND

Figure 2.4—Phase I Initial Insights (2-2)

Phase I: Initial Insights (continued)

Third, the internal structure of the TEPs inhibits the Joint Staff's ability to link objectives to resources, which is the traditional way resources are identified and allocated. No global objectives or uniform measures of effectiveness exist by which to identify what is needed and determine the appropriate resources required. The artificial separation of engagement from the other CINC missions precludes any rational association for the setting of engagement priorities and their resources. The level of detail found in the activity annexes of the TEPs are at too fine a level for OSD to adjudicate and make resource decisions.

Finally, the engagement mission's resources are provided in part by programs resourced by other federal agencies. Frequently, these resources are useful to supporting the mission but are not easily assured, since most non-DoD executive agencies do not use the DoD's planning and programming processes. Most agencies, such as the State and Justice Departments, operate on different schedules and time frames and are focused only on annual budgets and execution. The process and decision disconnects between other federal agencies and the DoD cause uncertainty within the DoD over resource availability and delivery to support the engagement mission. These disconnects in the processes of DoD and federal agencies create difficulties for the CINCs in both planning for and executing the activities involved with their theater engagement mission.

- **Viewpoint - Establishing the Demand**
  - **OSD - Responsible for strategic guidance**
    - **Engagement is part of National Security Strategy**
    - **Mission is not supported with specific programming guidance to services in DPG**
    - **PROs are too broad, separated from other CINC guidance**
  - **CINCs - Establish engagement mission objectives (operational demand)**
    - **Engagement is a continuing, all-inclusive mission**
    - **Regional and subregional objectives are specified in TEPs**
    - **Some have insufficient resources to execute TEPs**
      - **Forces are inadequate to support requirements**
      - **DoD resources are difficult to obtain**
      - **Interagency resources are uncertain**
    - **Risks to other assigned missions are not identified**

NDRI RAND

Figure 2.5—Viewpoint vs. Vantage Point (1-3)

Viewpoint vs. Vantage Point

Next, we discuss in several charts the perspectives of the players concerning some of the complexities of planning and resourcing the engagement mission. These insights were gained from preliminary interviews conducted with selected members of the OSD, Joint Staff, and service planning and programming staffs. Further refinements followed the discussions at the August 17, 2000, PWG Engagement Conference.

The OSD finds the engagement mission both complex and frustrating. The OSD is responsible for setting the strategic guidance for the DoD, and engagement is clearly part of the Defense Strategy. The engagement mission was formalized in 1997 with the establishment of the PROs in the CPG. The PROs were designed to provide broad strategic theater guidance to the CINCs and were never intended to be the foundation for resourcing the mission. The PROs are therefore responded to outside the DoD processes used to plan and program for other CINC activities. This shortcoming yields an engagement mission that lacks specific programming guidance in the DPG to the services in the PPBS process.

For example, the DPG for FYs 02–07 specifically mentions the engagement mission in its planning section. However, the critical programming section of the same DPG offers no guidance on what resources are to be provided for engagement, and no mention is made of engagement-related resource issues that the services must address in their programs.[8]

The CINCs also face numerous problems in identifying what is required to meet mission demands. For example, because engagement is a continuing inclusive mission, it uses capabilities provided for and associated with other CINC missions, but the quantities and frequency of resource utilization are not specified. In part, this is attributable to the development of the TEPs around regional, subregional, and country objectives. These geostrategic objectives do not inherently link to or identify types of operational capabilities needed to conduct engagement activities.

Some CINCs also argue that there are insufficient resources to fully execute the broad objectives laid out in the PROs for their TEPs. Given the high demand for capabilities to meet contingency demands and sustain readiness, it is impossible to provide sufficient resources for the engagement mission. In addition, some CINCs, such as the U.S. Southern Command (USSOUTHCOM) and the U.S. Central Command (USCENTCOM), contend that because they have few forces assigned to them, there are insufficient resources to conduct the engagement mission as laid out in their TEPs. The JSCP, which is part of the JSPS and lays out the assignment and apportionment of forces and capabilities to the regional CINCs, states in its section on engagement that engagement activities provide no basis for the assignment or allocation of forces.[9] However, the CINC needs to have forces involved to conduct combined exercises with the forces of other countries; lacking assigned forces constrains this major set of engagement activities.

These issues are further exacerbated for USSOUTHCOM and USCENTCOM given that many of their engagement activities are often dependent on resources provided by federal agencies outside the DoD. Frequently, their key engagement projects require long-term commitments to nations, and some of these projects include infrastructure development such as road construction. Without prior commitment of appropriate resources, it is difficult to integrate the several sets of capabilities required for such efforts. Often these activities are wholly or partially funded by other federal agencies outside the DoD. The complexities involved in getting sufficient, timely, and often multiyear

---

[8]U.S. Department of Defense, *Defense Planning Guidance, Fiscal Years 2002–2007* (U), April 2000.

[9]CJCSM 3110.01A, December 1998.

funding guaranteed by these other federal agencies frequently inhibit the CINCs' ability to adequately plan and execute their engagement mission.[10]

Because TEPs are developed within a separate planning stovepipe, they limit the CINC's ability to assess the impact of the engagement mission on the command's ability to perform the full span of its missions. Inasmuch as engagement activities are not strongly linked to the other CINC missions, the risk or impact on readiness of failing to execute engagement activities is not easily assessed.

---

[10]U.S. CENTCOM working papers, August 17, 2000.

- Vantage Point - Establishing the Supply
  - Services - Focus on DPG's programming guidance
    - Support specific executive agency engagement programs (e.g., Army for EUCOM Marshall Center)
    - TEPs rarely inform resource needs in service POMs
    - IPLs seldom raise engagement resource issues
    - Engagement is not a major program review or budget issue
    - Army and Air Force have own International Affairs Plans
  - Service Component Commands - Provide no direct input on engagement resource requirements to inform service POMs
    - Balance CINC guidance with service Title X responsibilities
    - Allocate assigned forces and resources to engagement without specifying tradeoffs (e.g., training OPTEMPO)

NDRI RAND

Figure 2.6—Viewpoint vs. Vantage Point (2-3)

Viewpoint vs. Vantage Point (continued)

As the providers of capabilities, the services operate as key players in the engagement mission. The services operate in two capacities in their provider roles. The first is executive agency responsibility for programs that involve more than one service. For example, the Marshall Center in Europe involves United States European Command (USEUCOM) joint service activities, but the Army has the executive agent responsibility for providing the overall resources (Figure 2.6).

The second is that the services develop their programs in response to the guidance and priorities established in the DPG. The development of the programming sections of the DPG is an iterative process in which the services actively participate. Since the engagement resource requirements are developed in a separate planning stovepipe, they get little attention from the services either in their review of the programming sections in the draft DPG or in the development of their respective programs. The services do provide significant resources to the component commands assigned to each of the CINCs. The majority of these resources are provided to support training and operational tempo (OPTEMPO) in the form of operations and maintenance funds for assigned forces. Since the capabilities provided by the forces and resources can be tasked by their respective CINCs for use in engagement activities, the services rarely receive direct requests from their component commands for engagement resources. Further, the services have little visibility on how their resources and assigned capabilities are employed for engagement by their component commands. Inasmuch as most of the resources provided to the CINCs in this way are fungible for engagement but are not seen as a direct programmatic requirement for the services, engagement rarely becomes a major program or budget issue when the services' Program Objective Memorandums (POMs) are reviewed by the OSD.

Our initial research indicated that the service component commands, which provide the immediate resources to support the regional CINCs' activities, do not always inform the service headquarters about engagement requirements and resources. Although this issue will be explored more fully in Phase II, it appears that the service components are not providing timely and sufficiently detailed data to their respective services in order to influence the development of service POMs and subsequent resource allocation. If a particular component supports an engagement activity such as training and exercise programs, the effects on their operational readiness and what activities they did not perform in order to support the CINCs' engagement requirements are not always explicitly articulated to either the service or the CINC.

The services have not been without involvement in the engagement mission and its resource requirements. The Navy provides operationally ready capabilities to the CINCs in the form of naval and marine forces within the Global Naval Force Presence Policy (GNFPP). The GNFPP allows the CINCs to use these temporarily assigned capabilities for engagement or other missions within their assigned time frames. The Army and Air Force generally provide assigned forces but must assure their training and readiness for employment, which limits their availability for engagement missions. Additionally, the Army and the Air Force are attempting to address the mission and its associated resources through the development of service International Affairs Plans (IAPs). The dilemma, however, is that currently these plans are not being developed in conjunction or coordination with the CINCs, the OSD, or the Joint Staff. Rather, they are internal service plans that define the engagement mission's requirements from an individual service perspective, focusing on security assistance and executive agency requirements with the associated resources to support them.

- **Viewpoint–Vantage Point - Integration**
  - **The CJCS and Joint Staff have little or no basis for integrating the fiscally constrained supply with non–fiscally constrained engagement demand**
    - **Global engagement is more complex than the sum of PROs or TEPs, which are outputs of unconstrained planning**
    - **No basis for determining the full measure of engagement requirements—total demand or what they achieve**
    - **Engagement activities lack relative priority; demand often operates at a microlevel of detail with no contextual framework**
    - **Key shortfalls and gaps in engagement activities are not identified and risks not assessed**
    - **Difficult to determine resource needs for engagement relative to other missions; unique mission without unique resources**
    - **Lack of useful measures of effectiveness for assessment (e.g., how many military-to-military contact man-days meet objective?)**
    - **No basis for integration of engagement with other missions**

NDRI RAND

Figure 2.7—Viewpoint vs. Vantage Point (3-3)

Viewpoint vs. Vantage Point (continued)

The CJCS and Joint Staff play an important role in the integration of the global engagement mission (Figure 2.7). The J-5 is responsible for the integration of the totality of the demand being placed on the CINCs for all missions; therefore, it also holds responsibility for integrating the global engagement mission. The outputs of the integration function are used as a basis for the apportionment and assignment of forces and capabilities in the JSCP. The J-5 also sets the demand for global contingency operations. The demand is not fiscally constrained.[11]

The J-8 is responsible for understanding the total supply of capabilities and resources that are being requested and for ensuring that it integrates and provides options for how the non–fiscally constrained demand can be supplied by the fiscally constrained resources.

[11]Chairman, Joint Chiefs of Staff, Instruction, *Responsibilities for the Management and Review of Theater Engagement Plans*, CJCSI 3113.01, Enclosures B and C, April 1998.

The dilemma for the Joint Staff is that the engagement mission, given how it is currently planned and structured, provides little or no basis for the integration of the fiscally constrained supply against the non–fiscally constrained demand from the several regional CINCs. Although more work will be done on this issue in Phase II, it appears that the global engagement mission is significantly more complex than simply resourcing the sum total of activities cited in the CINCs' TEPs. The effects of engagement activities on a command's components and readiness must be well understood prior to the initiation of engagement activities. The ability to do this is complicated by the lack of a global definition and objectives that are integral to the planning and programming of other CINC missions. For example, Operation Northern Watch (ONW), conducted by USEUCOM, is not considered or counted as an engagement activity; by contrast, Operation Southern Watch (OSW), conducted by USCENTCOM, is counted as both an engagement and an operational mission. These conflicting views need further synthesis and a common framework to ensure that a useful assessment of global engagement can be possible.

All of these problems inhibit the ability of the J-5 and J-8 to have visibility into the total requirements of the engagement demand or the total supply of resources available for the mission. The absence of a common framework for setting priorities across engagement and in relation to other CINC missions is an added impediment to the effective integration of resource allocation. The lack of these capabilities also precludes the development of useful measures of effectiveness for understanding how the engagement mission affects the broader set of CINC missions.

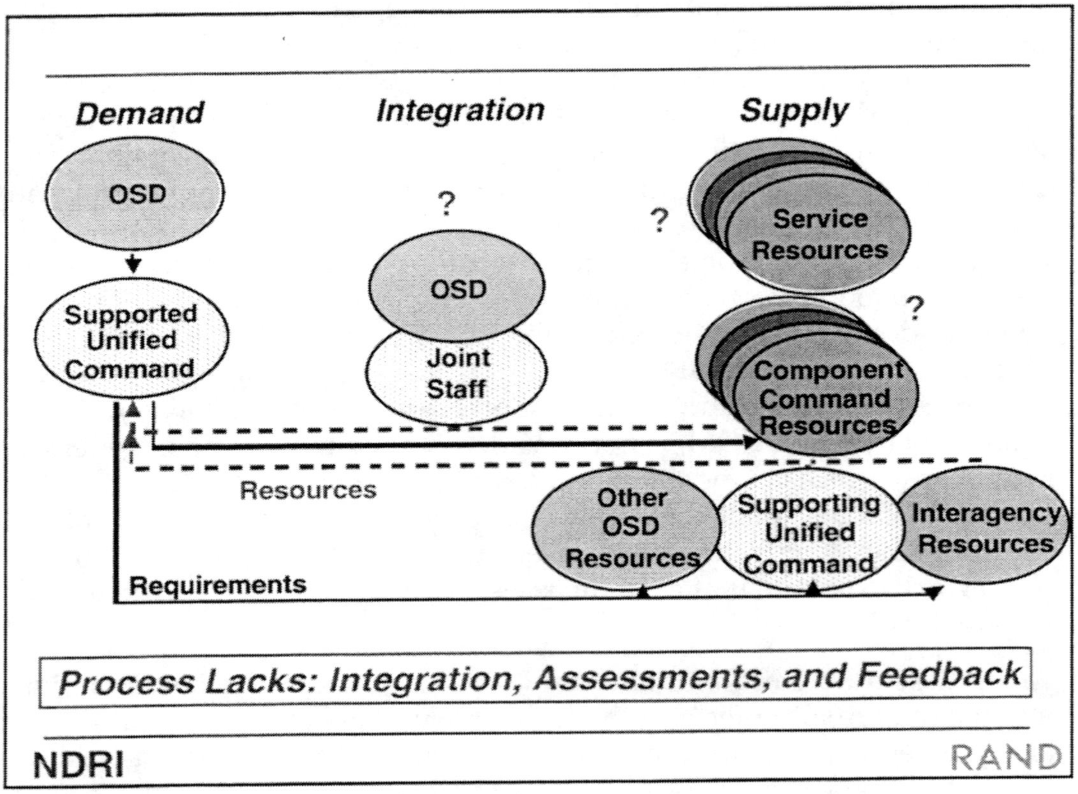

Figure 2.8—Theater Engagement Requirements and Resource Linkages

Theater Engagement Requirements and Resource Linkages

Using the economic model of supply, demand, and integration, the project team laid out how each of the players currently interacts in the planning and resourcing of requirements for the engagement mission. As indicated on the left-hand side of Figure 2.8, the demand is established at two levels. First, the OSD sets in the PROs the strategic guidance for what engagement objectives the CINCs must support with their activities. The supported CINCs establish their strategic concepts for engagement and initiate the demand requirements for resources to support their theater engagement activities. The requirements are articulated to the component commands, to supporting CINCs, and to the other federal agencies that support particular activities within a given theater.

Figure 2.8 shows that a number of key players and associated interactions are not taking place; the functional and process disconnects inhibit the ability of the DoD to understand the complexity of the mission. The right-hand side of Figure 2.8 shows that the services do not have clear linkages to the demands being placed on their components in terms of what is being asked of them to support the CINCs' theater engagement objectives. Similarly, the supporting unified commands are not tightly connected to the process except in providing resources requested by the supported CINC. There are no feedback loops concerning how successful the capabilities provided by the components or supporting CINCs were in meeting the engagement activities. The dotted line in Figure 2.8 identifies problems related to coordination and interaction with the other federal agencies for their resources; the question marks indicate a lack of needed interaction.

The OSD and Joint Staff are not connected to the determination of the total demand (CINC requirements), nor do they have visibility into the total supply of resources available to support the engagement mission (see the center of Figure 2.8). This lack of connectivity among the players prohibits the Joint Staff and the OSD from doing their job of integrating global demand and supply as well as adjudicating the differences between the unconstrained demand for engagement resources against the finite supply. Because of this problem, the J-8 really concentrates on identifying and attempting to fill the gaps and shortfalls in engagement resources as they are identified by the individual CINCs. This narrowly focused process prevents the J-8 from gaining an understanding of the relative priority of various engagement activities and their associated resource requirements. Hence, global engagement mission requirements and needed resources cannot be objectively assessed.

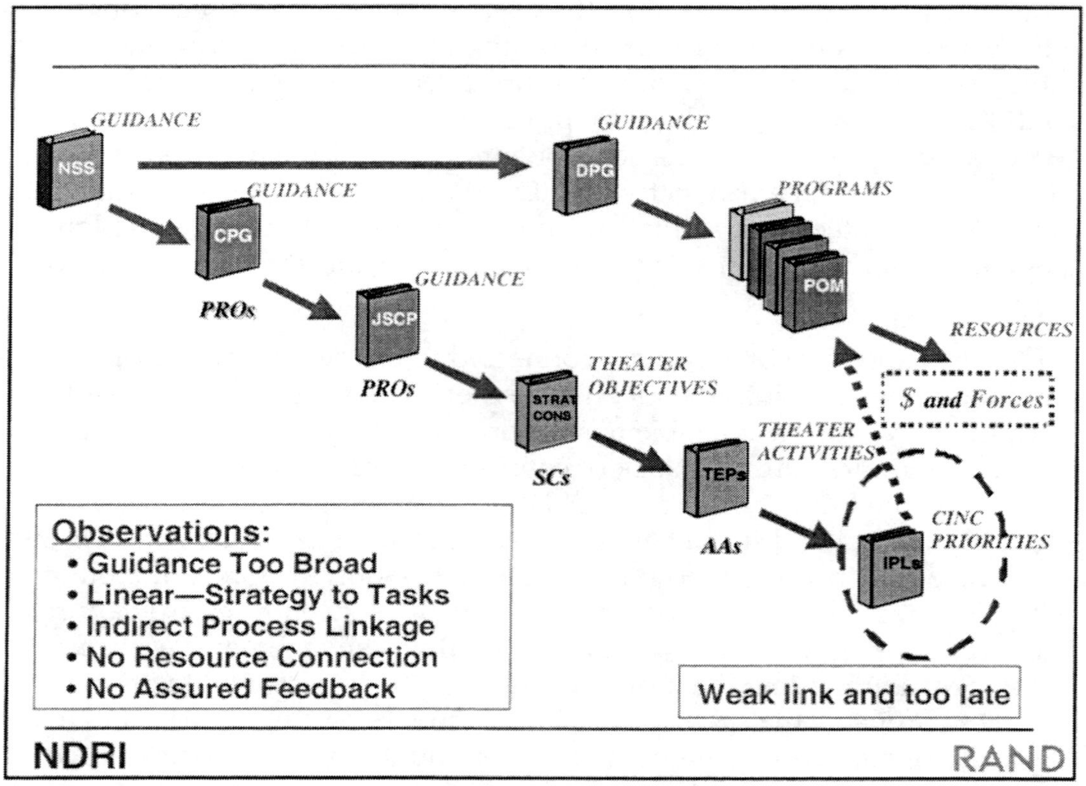

Figure 2.9—Engagement Process in Perspective

Engagement Process in Perspective

Another way to illuminate the planning and programming problems associated with the global engagement mission is to examine how its requirements are identified and linked to the JSPS and PPBS processes. Key aspects of these processes are illustrated in Figure 2.9.[12]

The diagonal arrows extending from the upper left-hand side of Figure 2.9 downward show the current TEP process and its linkages to the PPBS process. The engagement mission is based on the National Security Strategy (NSS) and its emphasis is on shaping as one of the three strategy pillars. The specifics of the mission are contained in the CPG. The CPG, however, does not articulate the specific goals and objectives of the mission; rather, it directs the CINCs to develop TEPs based on the PROs assigned to their region. The JSCP is the deliberate planning document

[12]CJCSI 8501.01, Enclosure B, April 1999, and CJCSM 3113.01A, Enclosures A and B, April 2000.

that is used to apportion and assign missions and forces based on the military strategy; however, it indicates that engagement activities do not provide a basis for the allocation of forces or other resources. Instead, it again directs the CINCs to lay out a TEP based on the PROs.[13]

The TEPs provide the first place in which specific theater objectives are identified and associated with resources. Resources are identified in the activity annexes of each theater's TEP. Although the TEPs are (by CJCS directive) developed according to a strategy-to-tasks structure, they are inconsistent in their details from theater to theater. For example, USEUCOM's TEP is highly detailed and is based on regional, subregional, and country objectives that are associated with approximately 5000 annual activities. By contrast, USCENTCOM's TEP is based on 15 strategic theater objectives, five of which relate to engagement. USCENTCOM's TEP strategic concept is approximately 15 pages long, and its activity annex contains fewer than 100 critical annual activities, whereas USEUCOM's TEP is multivolumed.

The Integrated Priority Lists (IPLs) are the most important direct entry point into service programs for theater engagement. Within the broader PPBS process, however, the IPLs' principal purpose is to identify critical CINC resource gaps or shortfalls, not to inform the programming guidance in the DPG that initiates the services' program development. Although the services review and give consideration to the IPLs, which provide a direct mechanism for the CINCs to identify selected needs to be adjudicated in the service programs, the IPLs are generally considered exceptions after most of the service programs have been built. Initiatives are under way in the Joint Staff and services to make sure that the IPLs enter the system before the service programs are built. However, the IPLs currently provide a weak link compared to program guidance contained in the DPG for receiving resource consideration by the services in the PPBS process.[14] Engagement resource needs should be considered earlier in the PPBS cycle and IPLs used only for exceptional shortfalls.

More telling than the impact that IPLs might have in resourcing the engagement mission is the fact that planning for the engagement mission operates in a stovepipe distinct and separate from the formal ways the DoD plans, programs, sets priorities, and allocates resources. The right-hand side of Figure 2.9 identifies the more formal process. The DPG is the mechanism that provides programmatic guidance to the services. The JSCP provides guidance, allocates forces, assigns missions, and sets priorities for the CINCs. A service allocates resources in response to the

---

[13]U.S. Department of Defense, *Contingency Planning Guidance, FY 1999–2003* (U), January 1998, and CJCSM 3110.01A, December 1998.

[14]CJCSI 8501.01, Enclosure B, April 1999.

DPG. The service then attempts to balance the multiple demands placed on it with the available resources. The adjudication of the service programs occurs in a series of hierarchically managed exchanges between the OSD, Joint Staff, and services. At the end of the process, the SecDef directs the services to implement decisions that are presented in the DoD Future Years Defense Plan (FYDP) and subsequently sent to Congress for approval in the budget.

The RAND assessment determined that the engagement mission is operating in a separate planning stovepipe that is not well connected to the JSPS or PPBS processes. The linear approach used in the strategy-to-tasks-developed TEPs does not provide the clarity and linkages to existing programs within the regional commands, nor does it facilitate the identification of sets of critical capabilities associated with performing the engagement mission. The IPLs are too weak a link to the formal decisionmaking processes to ensure that engagement resource demands are visible within the broader PPBS structure. Finally, the lack of clear connections to resources inhibits the ability of the CINC and the OSD to understand how the engagement mission is being conducted, the effectiveness of engagement activities, and the utility of resources expended.

- **JSPS**
  - JSCP (Guidance, Priorities, & Apportionment)
  - JPD (Guidance & Priorities)
  - CPR (Priorities)
  - CPA (Resource Allocation)
  - IPLs (UFRs & Priorities)
  - TEP S/Cs (Guidance)
  - TEP A/As (Requirements / Resources)
- **PPBS**
  - DPG (Guidance & Priorities)
  - CPG (Guidance & Priorities)
  - POMs (Resource Allocation)
  - PDMs (Resource Allocation)
  - PBDs (Resource Allocation)
  - PB (Resource Allocation)
  - AA (FY Appropriation Apportionment)

Detailed assessment of process content and linkages—poor connectivity without visible integration

NDRI RAND

Figure 2.10—Potential Process Portals for Engagement Requirements and Resources

Potential Process Portals for Engagement Requirements and Resources

The critical documents associated with the two major planning and programming activities in the DoD were evaluated to determine where engagement activities are contained or supported. In Figure 2.10, the underscored abbreviations—JSCP, IPLs, TEP SCs, TEP AAs, and CPG—show where the engagement mission is identified and discussed in the current process. The remaining key documents are those that contain little or no mention of engagement and its resourcing but are potential portals for supporting engagement in the future. In the JSPS system, for example, two critical documents that articulate CINC demands and then assess service programs' responsiveness to these demands are the Chairman's Program Review (CPR) and the Chairman's Program Assessment (CPA). Other than the CPG, the identification and allocation of resources within the PPBS process in support of the engagement mission do not occur. Although it identifies the mission in the planning guidance, the DPG is not included as a TEP supporting document because it contains no program guidance for the engagement mission. Because of the lack of linkage to the formal decisionmaking processes within the DoD, the mission and its resource demands have little or no visibility when the service programs are reviewed and resource allocation is adjudicated.[15]

---

[15]CJCSI 3100.01A, September 1999, and CJCSI 8501.0, Enclosure B, April 1999.

Engagement

- **Current engagement process status—lack of resources is a symptom of more complex issues:**
  - **Guidance and priorities are too general and engagement is not related to other CINC missions**
  - **Planning process is immature but evolving—not fully integrated within the JSPS**
  - **Programming guidance does not address engagement**
  - **Requirements are isolated from the totality of resources**
  - **Little basis for global assessment or measure of effectiveness**
  - **Players and existing processes are not used effectively**
  - **Need to improve interagency support for resources**
- **Expect engagement mission to continue in any future strategy**

NDRI RAND

Figure 2.11—Initial Assessment of Engagement Problem

RAND Assessment of the Engagement Problem

The problems associated with the planning and programming of the engagement mission are summarized in Figure 2.11. The guidance and priorities in the PROs are too broad and general to support resource allocation. The JSCP guidance for engagement provides no integration with other missions and the TEP process is not fully integrated within the JSPS. Although the OSD has realized this and is attempting to address some of these problems through changing aspects of the PROs, the engagement mission remains largely disconnected from the major DoD decisionmaking processes. The OSD continues, however, to work on reshaping the guidance and has agreed to work with the project team in Phase II to improve the PROs and other aspects of the engagement problem as they are affected by the PPBS process.

Another problem is that the programming guidance in the DPG does not address engagement. This shortcoming must be redressed in order to ensure that the mission is seriously considered in the broader resourcing environment. How best to provide programming guidance will be a key element of the Phase II study.[16]

Certainly coordination and ensuring that resources are received from the other federal agencies is critical. However, all players recognize this problem as a difficult one that is not within the prerogatives of the OSD to remedy. Phase II will attempt to identify some improvements that will address this area.

Regardless of the future administration transitioning in 2001, it seems clear to all players involved that the engagement mission will continue to be a critical element of future defense strategy. As such, the recommendations developed in Phase II of the study may support the normal evolution of change that accompanies the transition of administrations and the upcoming Quadrennial Defense Review (QDR) 2001.

---

[16]U.S. Department of Defense, *Defense Planning Guidance, Fiscal Years 2002–2007* (U), Planning and Programming Guidance Sections; April 2000.

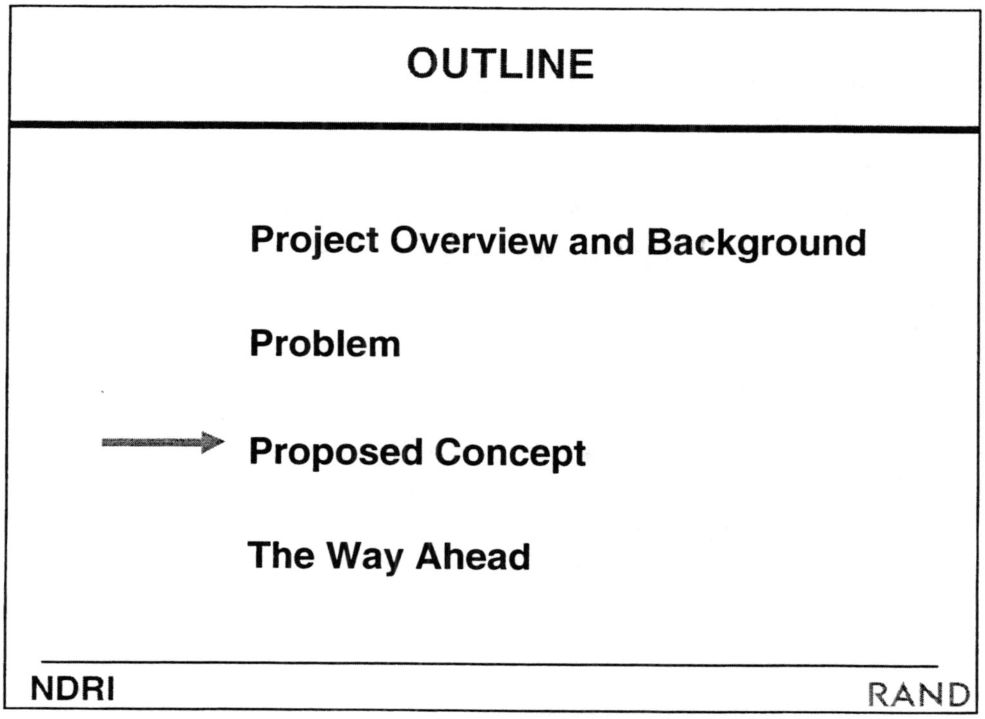

Figure 3.1—Outline

PROPOSED APPROACH FOR DEVELOPING ENGAGEMENT IMPROVEMENTS

Outline

This section lays out the proposed concept for developing improvements to the engagement mission that will be further developed in Phase II of the study (Figure 3.1).

- **Develop a basis for global engagement integration and assessment**
- **Develop a hierarchical approach that links all the players and processes at the appropriate levels**
- **Eliminate engagement stovepipes by systematically incorporating engagement into the existing JSPS and PPBS**
  - **DPG must provide a critical link with specific guidance to services on programming for engagement**
  - **Fully mature and integrate TEPs into JSPS**
  - **Use formal PPBS decision process inputs and feedback loops**
- **Structure engagement and its resources within the context of operational missions and readiness—accepted and understood**
  - **Provide a basis for global engagement assessments**
  - **Develop an improved context for engagement resources**

NDRI RAND

Figure 3.2—Attributes of the Concept

Attributes of the Concept

Any new concept for improving engagement planning and programming needs to address the key problems identified in the current process (Figure 3.2). The focus of the project must include engagement from a global perspective, and therefore any proposed concept must define a hierarchical structure that disciplines the system while allowing all the participants—CINCs, services, Joint Staff, and the OSD—to interact at the appropriate levels. The concept must ensure that engagement is properly embedded in existing DoD decisionmaking processes (e.g., JSPS and PPBS); therefore, the concept must not only show how engagement could be linked but also explicitly eliminate the functional and planning stovepipes in which that mission is currently operating. The engagement activity thus needs to be explicitly defined so that it can be discussed and understood on a common basis within both the planning and programming functions as they operate in the JSPS and PPBS.

This is not to suggest that TEPs need to be eliminated or radically altered; rather, they need to be sufficiently developed and integrated to inform the needs of both the JSPS and PPBS processes. If the TEPs could be further matured to provide this type of information, then engagement could be tied into the PPBS and its hierarchical decisionmaking processes to improve the potential for resource allocation.

Based on initial insights, it became clear to the project team that engagement is another CINC mission that has yet to be fully formalized in the DoD planning and programming structure. It is different, however, in that it operates as an enabler of other CINC missions. Placing engagement within this broader operational context appears to facilitate an improved basis for assessment of the mission and its ability to better compete for scarce resources.

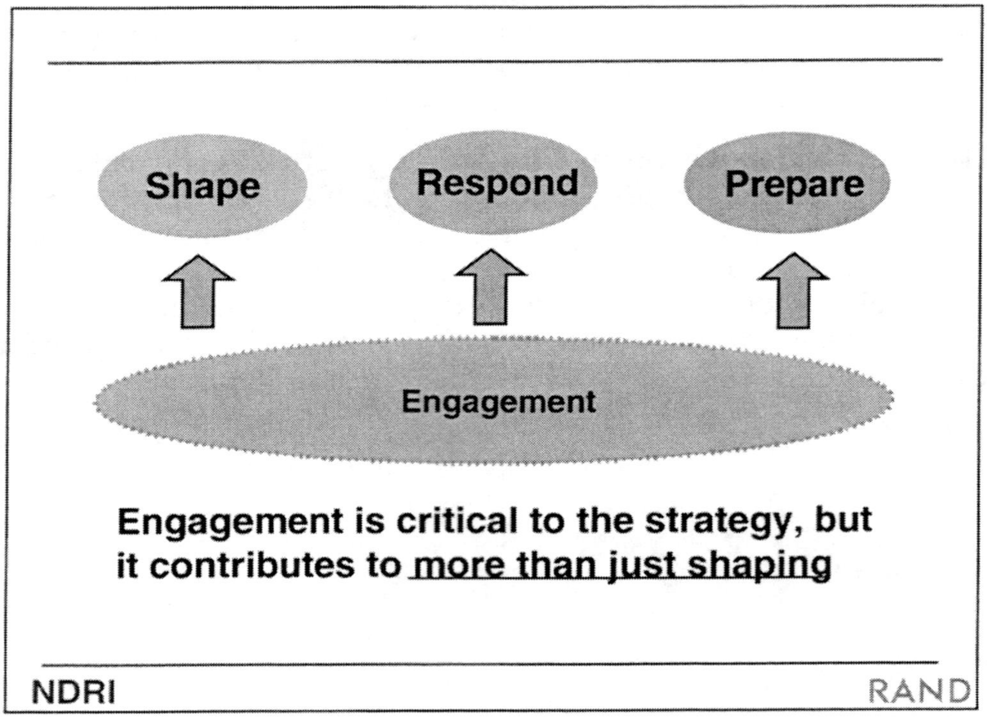

Figure 3.3—The Emerging Engagement Paradigm

Emerging Engagement Paradigm

Engagement should not be restricted to the shape portion of the Defense Strategy. Rather, the engagement mission seems more appropriately viewed as a critical enabler of all three pillars of the Defense Strategy. Figure 3.3 illustrates some of the initial rethinking of how engagement supports the entire Defense Strategy.

The TEPs were initially conceived as part of a strategic planning function in that the PROs were viewed as ways in which the OSD could provide some top-level planning guidance to the CINCs solely for shaping the peacetime environment. The original scheme for engagement was not viewed as a mission that demanded additional resources or support; rather, it was attempting to provide some structure to activities that were already occurring within the regional CINCs' theaters. Although the demands of the engagement mission have become better understood, the resource tail has been largely dependent on resources already allocated for the other traditional CINC missions. Broadening the mission's definition to extend across all three pillars of the Defense Strategy ensures that engagement has the increased visibility and emphasis that it needs to be seen as a critical enabler of operational readiness in the demanding multimission operational environment in which the CINCs operate.

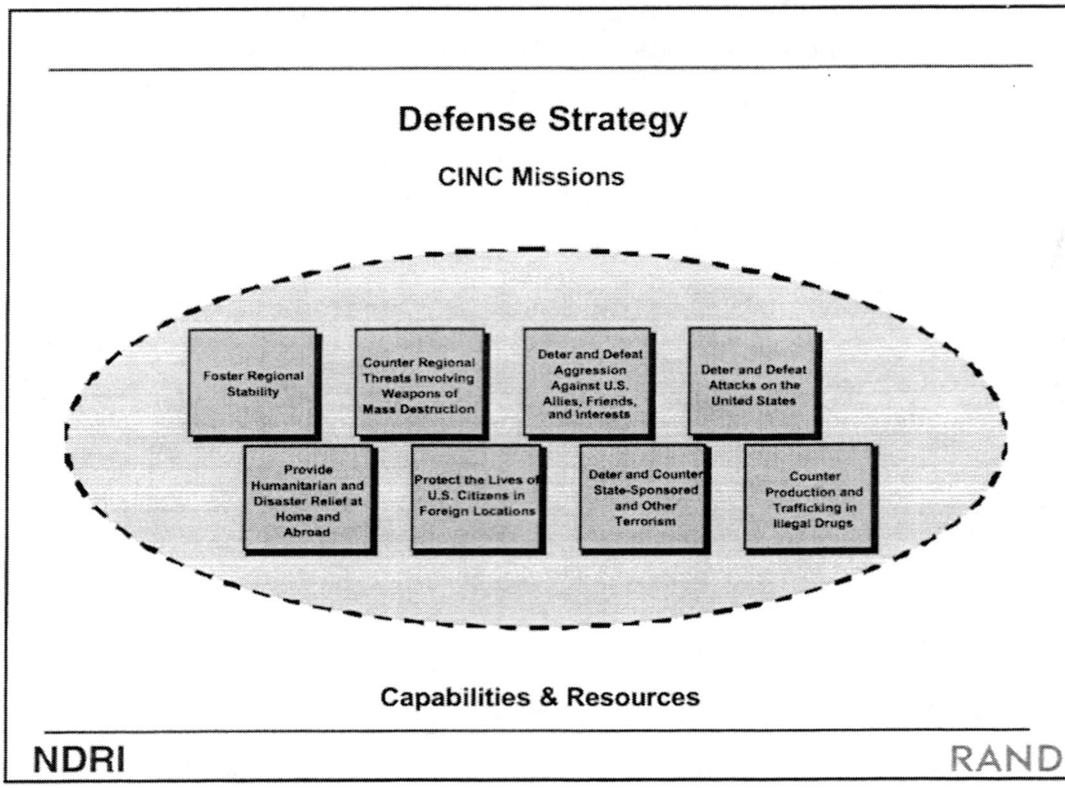

Figure 3.4—Applying the Paradigm

Applying the Paradigm

Viewing engagement as a major contributor to all three pillars of the Defense Strategy is critical to providing a broader contextual framework for that mission.

Engagement appears as an enabler of the traditional CINC operational missions depicted in Figure 3.4 and as a key contributor to operational readiness. Linking engagement to these operational missions provides an improved contextual framework within which engagement requirements can be assessed for both priority and risk. It also appears to ensure that the resources required for engagement activities are better understood in terms of their contribution to all three pillars of the Defense Strategy—shaping the strategic environment, and responding to and preparing for future contingencies. Another dimension of this approach is that CINCs could provide some direct linkages to the resource demands of the engagement mission as they relate to or affect the other missions in their theater.

This approach has the potential to strengthen the linkage to resource allocation—the CINCs and services will be able to make more compelling arguments for resources in the broader operational mission context.[17]

---

[17]See Leslie Lewis, James A. Coggin, and C. Robert Roll, *The United States Special Operations Command Resource Management Process*, for an earlier application of this approach.

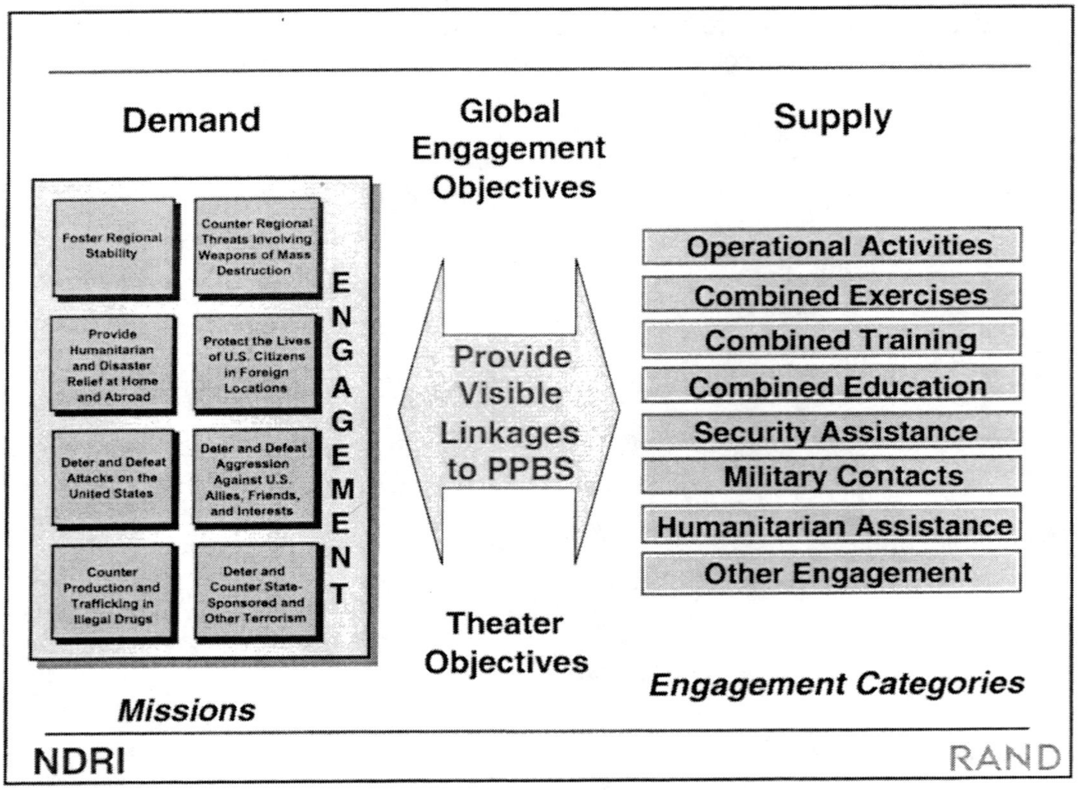

Figure 3.5—Linking Engagement Activities to Support CINCs' Operational Missions

Linking Engagement Activities to Support CINCs' Operational Missions

The key attribute of the proposed approach lies in the assumption that engagement is not merely an ancillary mission performed by CINCs during periods of peace but rather a key contributor to implementation of the Defense Strategy. Engagement is also critical in ensuring that the United States and its allies are ready to meet any operational demand. Figure 3.5 shows how the new conceptual approach might be viewed with the engagement activities enabling the traditional CINC missions.

Applying the model of supply, demand, and integration, the engagement mission is shown in the left-hand side of Figure 3.5 as contributing to the totality of CINC missions. It is the everyday things that CINCs do that ensure readiness; hence, it is vital that the mission be performed. The right-hand side of the figure (the supply side) shows the eight existing engagement categories of activities that are currently associated with the mission.[18] While some might debate the utility of having some operational activities and humanitarian assistance activities as separate categories on the supply side, the eight engagement categories represent those things that are currently planned and executed to prepare for the breadth of operational missions that the CINCs must perform. These issues will be more fully explored in Phase II.

---

[18]CJCSM 3113.01A, Enclosure C, April 2000.

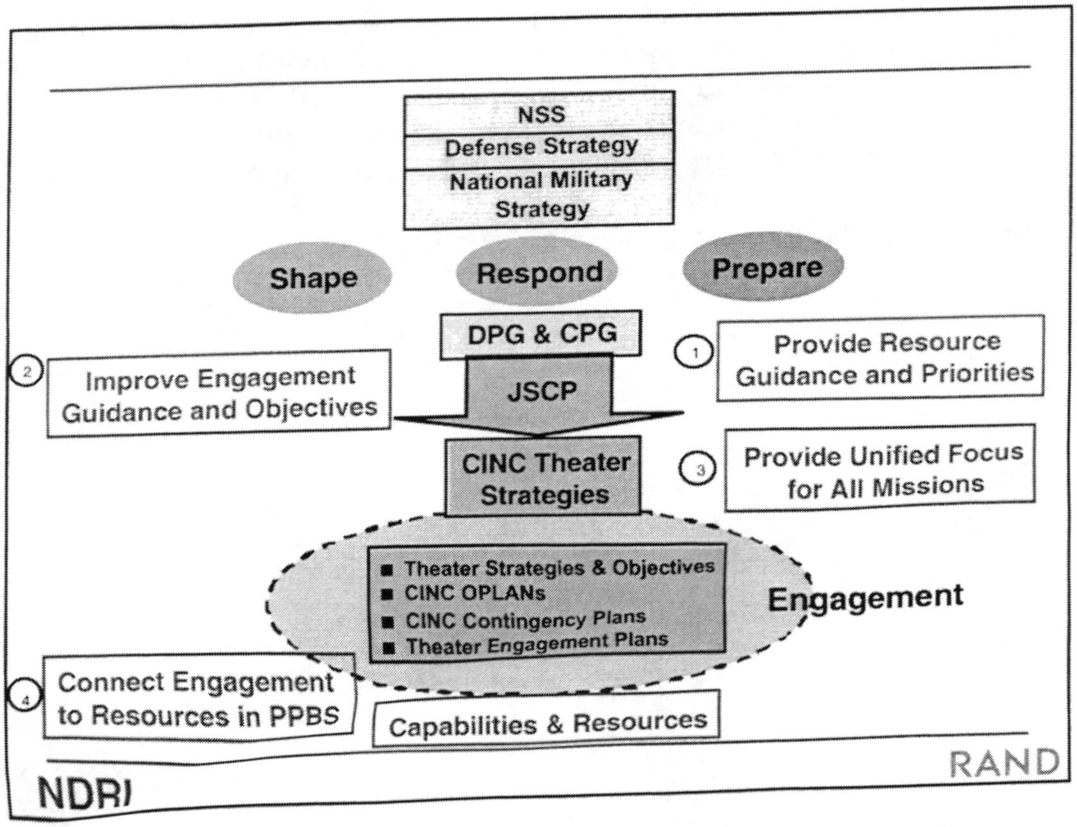

Figure 3.6—Concept for Improving Engagement's Visibility

Concept for Improving Engagement's Visibility

Figure 3.6 outlines the key attributes of this concept and some of the steps needed to address the *full* scope of the engagement mission. As noted earlier, this approach ties the engagement mission to other CINC missions. *It is* planned and programmed for within the existing structures for *all* other CINC missions.

The guidance and priorities provided in the CPG, DPG, and JSCP will need some reshaping and refocusing to ensure that engagement is fully integrated into the JSPS and PPBS processes. Engagement planning and programming activities will also need to be synchronized with the timing of formal decisionmaking processes to ensure that its demands are articulated, considered, and adjudicated.

CINC TEPs will no longer be just stand-alone documents linked only to the shaping activity; instead, they will need to become integrated enablers and part of the CINCs' Theater Strategy based on all guidance, not just the PROs. In this manner, engagement will be structured within the broader context of CINC operational missions and contributions to readiness. This approach will increase the potential for engagement to compete within the full spectrum of resources and for the OSD and Congress to understand the causal relationship among the engagement mission and other CINC operational missions.

Initial review and discussions of TEPs have revealed that most CINC staffs want these documents to remain useful planning tools within their individual commands even if this broader approach is not fully adopted. The issue of what role the TEPs should play in planning and programming for the engagement mission or how they may be adapted for these purposes will be fully addressed in Phase II.

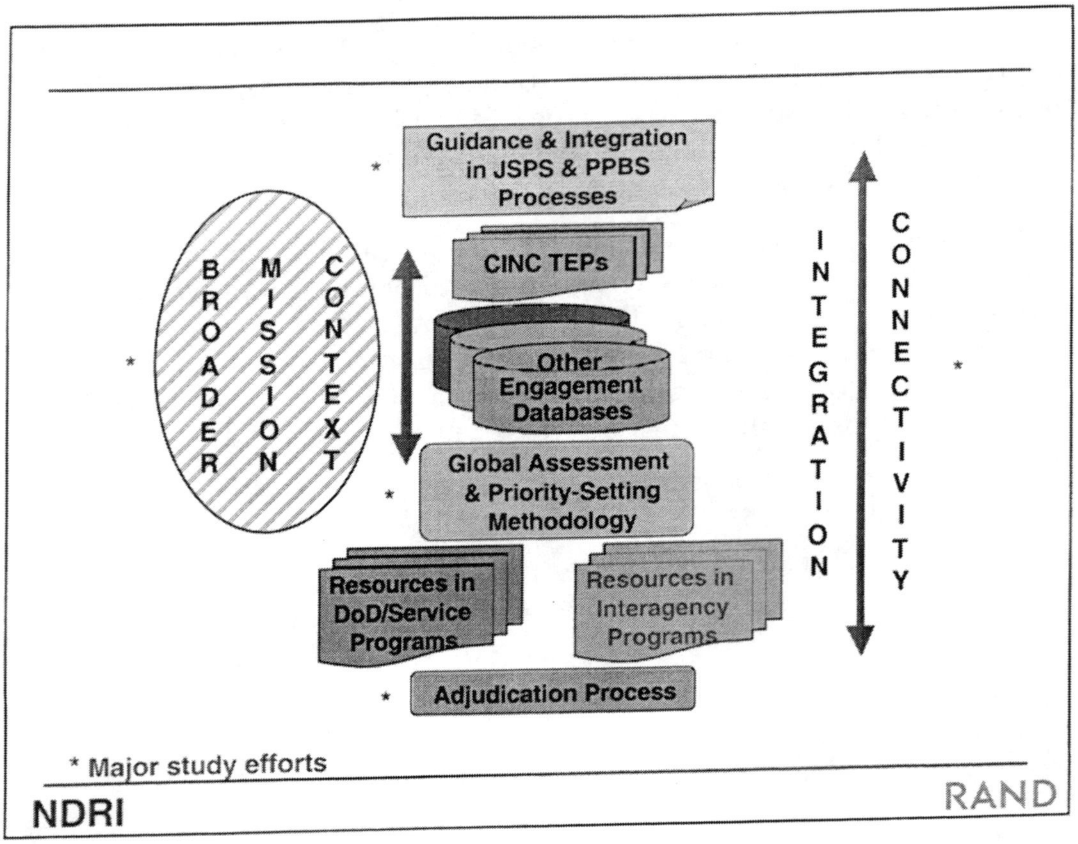

Concept for Improving Engagement

Figure 3.7 further illustrates the concept for improving our engagement efforts. Again, this is an initial concept that will be further developed in Phase II on the basis of detailed discussions and interactions with the CINCs and their staffs, the services, the Joint Staff, and the OSD. Those tasks in Figure 3.7 with asterisks next to them are key elements of the Phase II study. Placing the mission within the broader mission context is a key element of the proposed approach. We believe that this approach will ensure improved integration with and connectivity to the appropriate activities of the JSPS and PPBS.

The approach enables all the players—the OSD, CINCs, services, and Joint Staff—that are directly affected by this mission and its inherent parts to participate in a structured manner in identifying their requirements and providing alternatives for improvements. Critical tasks in the concept for Phase II work are the identification of ways to improve guidance and process integration; the development of measures for assessing the effectiveness of global engagement and setting priorities that will assist in adjudication of resource allocation; analysis of the sensitivity of their interrelations; and comparative evaluation of potential options for improvement.

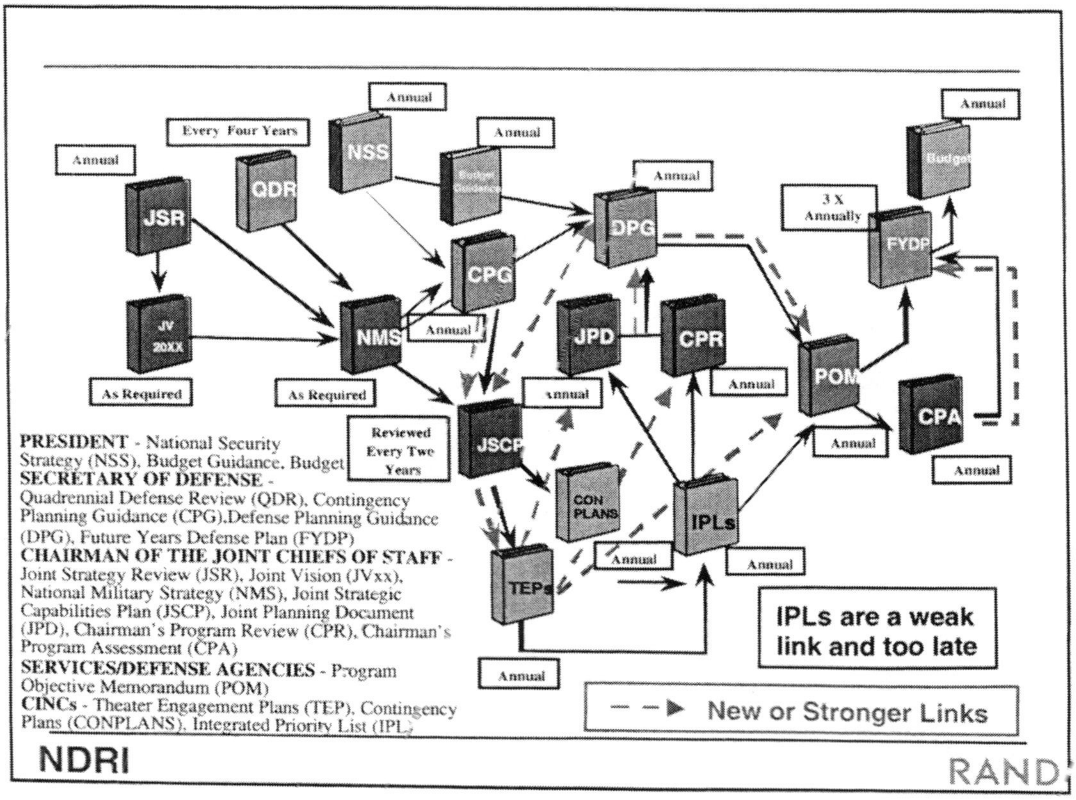

Figure 3.8—Integrating Engagement into JSPS and PPBS

Integrating Engagement into JSPS and PPBS

Figure 3.8 shows the key documents and activities in the JSPS and PPBS. The existing responsibilities of the OSD, Joint Staff, CINCs, and services in the development of the products are shown on the left-hand side of the chart. The arrows indicate the current flow of guidance and development through these major processes, culminating in the FYDP and annual budget.

Figure 3.8 postulates with dashed lines and arrows some potential uses of various documents and their linkages. Also shown is the timing that may need to be strengthened or reordered as well as the links that may need to be reengineered to ensure that engagement is fully connected and integrated with these major decisionmaking processes. In this way, the potential can be increased for obtaining engagement visibility and resources. This approach moves beyond the use of IPLs as the principal link for the engagement mission into the PPBS process. Rather than usir exceptional means to raise resource issues, the proposed changes woulc make engagement similar to how other CINC missions are planned for and resourced by linking engagement activities across all major activities associated with the formal JSPS and PPBS processes.[19]

[19]CJCSI 8501.01, Enclosure B, April 1999.

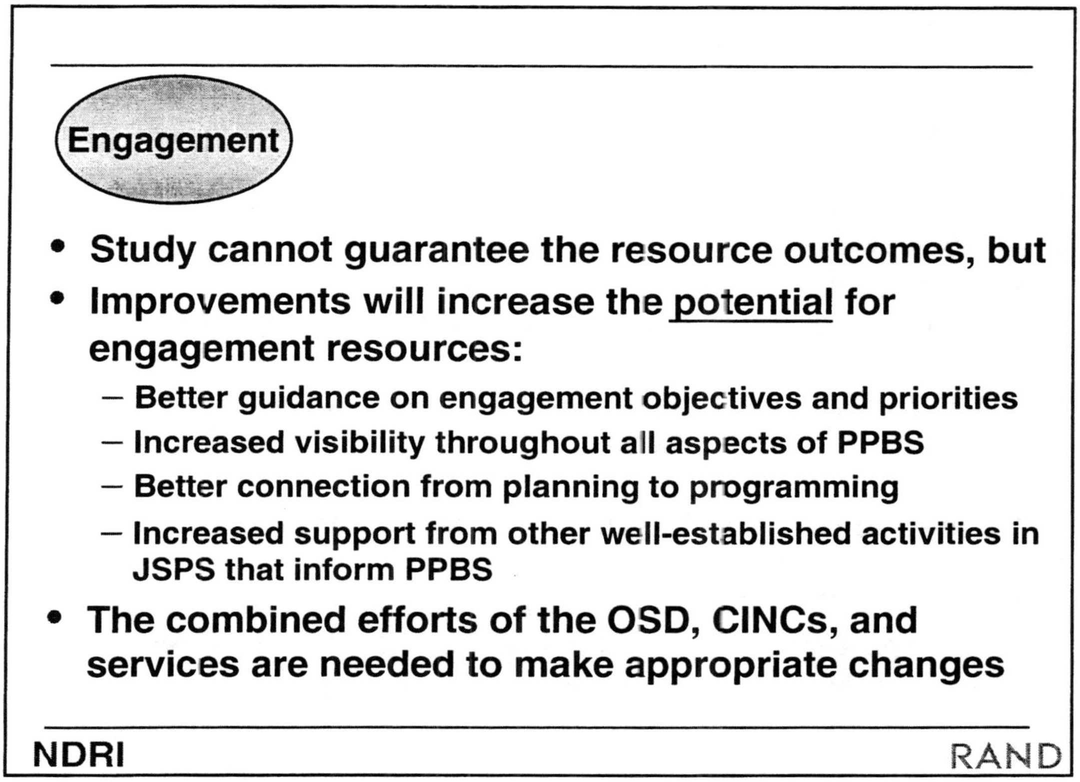

Figure 3.9—Some Observations

Some Observations

Clearly, the RAND project team and its sponsors cannot guarantee the final outcome for sustaining current or even additional resources for the engagement mission. It seems equally clear, however, that to increase the potential for resources, the engagement mission must move out of its current planning and programming stovepipe and be embedded within the established JSPS and PPBS processes. The formal decisionmaking processes provide well-understood mechanisms that facilitate the justification of the mission, the identification of required capabilities, and appropriate visibility for adjudicating resource allocation and funding. By connecting engagement activities to a broader set of CINC operational missions, we believe that realistic objectives can be developed that will yield requirements to compete for resources across the spectrum (Figure 3.9).

- Expect the engagement mission to continue in any future defense strategy
- Need to continue to evolve TEP—still immature; must get engagement out of current stovepipe
- Linking engagement to other operational missions provides a foundation for global assessment and demand for additional resources
- OSD and Joint Staff need to provide engagement-specific programming guidance, objectives, and resource priorities
- Need to use existing decisionmaking and resource allocation processes
- Phase II RAND study will provide the vehicle to connect and integrate engagement within JSPS and PPBS
- Need CINC support for evolving the process

NDRI RAND

Figure 3.10—Summary

Summary

Engagement is not a new mission but rather one that has gained prominence since the fall of the Soviet Union and the emergence of a multipolar strategic environment. Although the new administration will likely develop new national security and defense strategies in preparation for the upcoming QDR, it seems certain that the engagement mission will continue to be an essential element of any new defense strategy (Figure 3.10).

The TEP process is an immature but evolving process, and like any relatively new activity, it can be improved with refinement of purpose and feedback from use as it develops. The concern is that although the TEP provides a useful mechanism with which to capture the complex nature of the activities and resources associated with the engagement mission, it cannot operate to its full effectiveness in its current planning stovepipe. The TEP is not sufficiently linked to the formal DoD decisionmaking processes—JSPS and PPBS—to ensure that engagement activities are considered within the broader context of DoD planning and programming. In order to do this, the engagement process needs to build on the accepted DoD processes and products.

In Phase II, some of these initial concepts and ideas will be further fleshed out through interviews and review of CINC, OSD, Joint Staff, and service engagement activities. Critical to any successful options being proposed is to ensure that all viewpoints are obtained, understood, and considered.

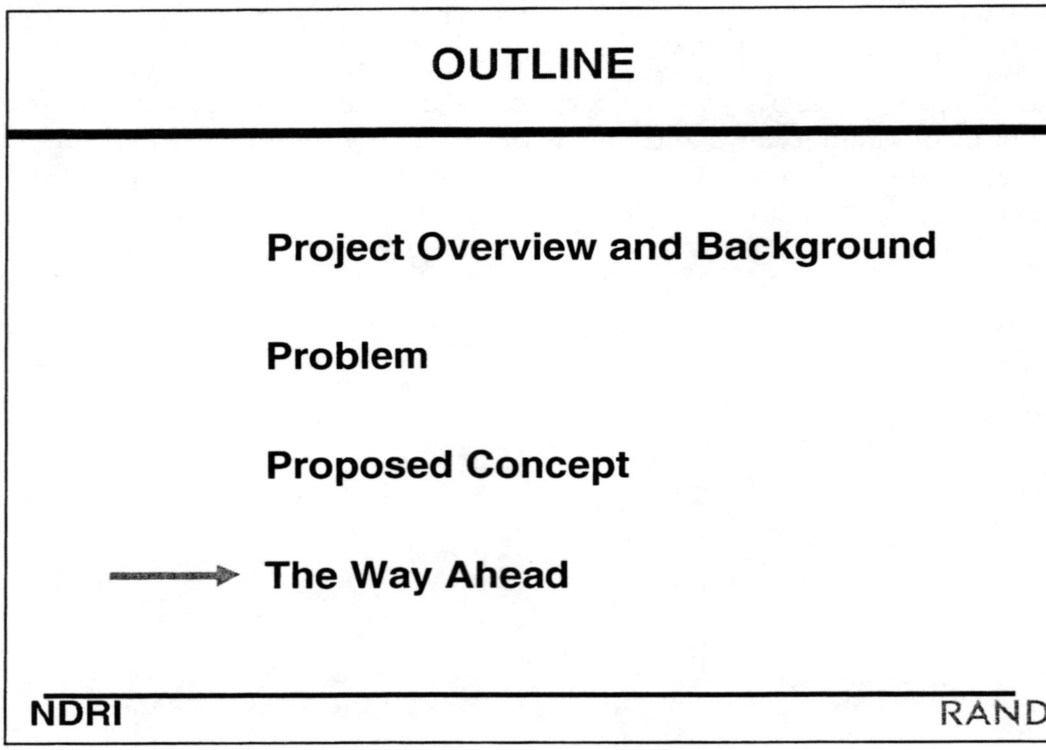

Figure 4.1—Outline

THE WAY AHEAD

Outline

This section provides the key activities in the way ahead and the tentative milestone schedule for our Phase II efforts (Figure 4.1).

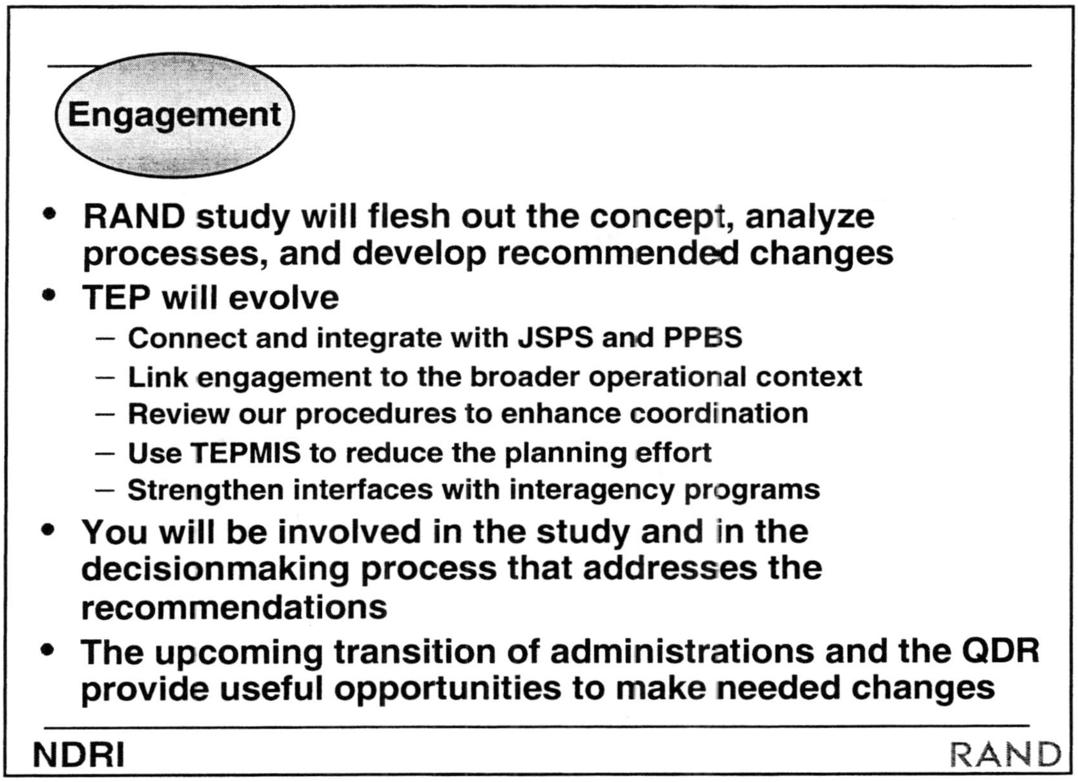

Figure 4.2—The Way Ahead

The Way Ahead

In Phase II of the study, the initial concept and ideas will be further developed and fleshed out by the research team through discussions and interviews with CINC, OSD, Joint Staff, and service engagement players. These efforts will be supported by a detailed review of the formal planning and programming processes and their specific activities as they may relate to the engagement mission (Figure 4.2).

The research team will work in close coordination with its study sponsors in the OSD and the Joint Staff, J-5 and J-8, to evolve the theater engagement mission by providing timely periodic recommendations for changes throughout the duration of the study. Specific tasks will include analysis of methods to fully integrate the engagement mission within the broader aspects of the JSPS and PPBS to include ways to improve both planning and programming guidance. In accordance with our concept of connecting engagement to the broader operational context of other traditional CINC missions, the research team will develop and examine alternative methods by which TEPs can be refocused to accomplish this objective while identifying ways to simplify and reduce the administrative burden of TEP development. Another task will be to review and evaluate ways to enhance the coordination of engagement requirements and capabilities and resources between supported and supporting players at the various levels. A further task will be to assess other federal agencies and their processes for providing input and resource to the TEPs and to determine the potential for improvements.

Throughout the study, it is critical that all major players in the engagement mission be kept fully informed as to the development of any options for improvements and be given the opportunity to provide feedback before decisions are reached. The project team, in coordination with the study sponsors, will make a major effort to ensure that all viewpoints on problems and potential solutions associated with engagement are obtained, have been fully understood, and are considered in the process of change.

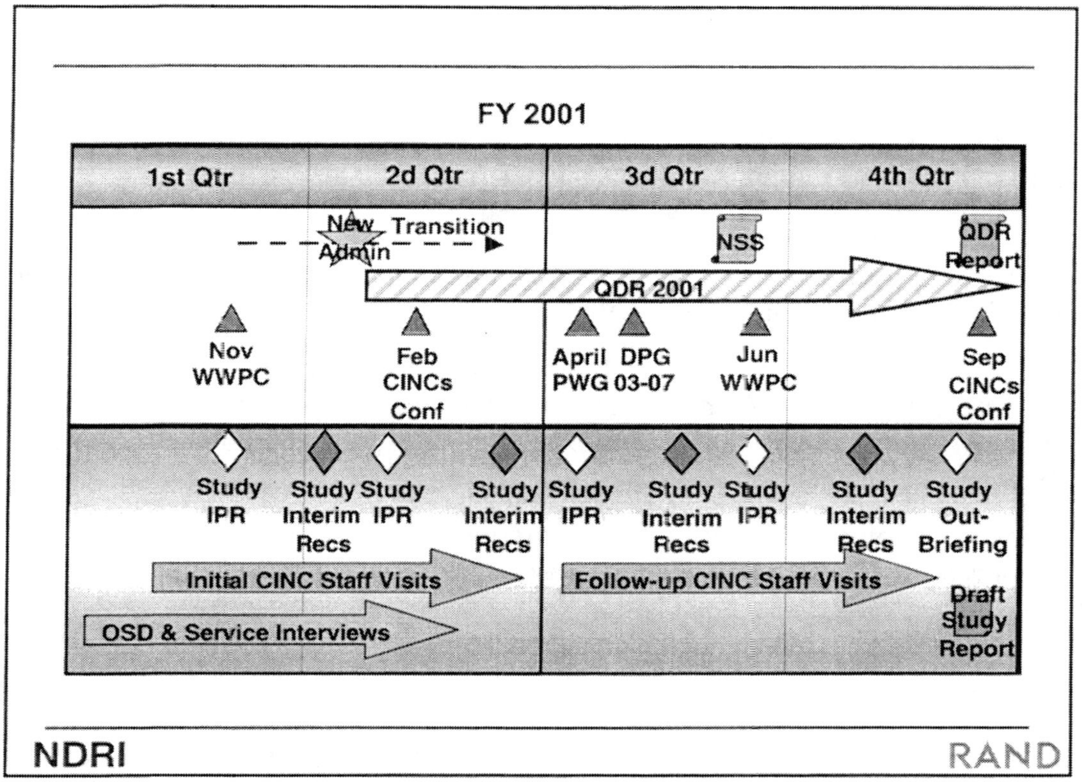

Figure 4.3—Proposed Study Milestones

Proposed Study Milestones

Figure 4.3 illustrates the tentative milestones for Phase II of the study. The top of the chart shows key concurrent activities and meetings with which the tasks of the study must be synchronized. The study of in-process reviews (IPRs) will be timed and scheduled to provide appropriate outputs to inform and support the key events within the JSPS and PPBS processes and significant meetings related to the engagement mission. As noted in the figure, the early portion of the study will provide time for research and collection of information from the major players in the engagement mission. This will include visits to the supported regional combatant commands and selected supporting CINCs. The study will culminate in a final briefing to the sponsors prior to the CINCs' Conference in September 2001. Subsequently, and before the end of fiscal year 2001, the project team will provide a draft report that describes the study effort and recommends steps to improve accomplishment of the engagement mission.

BIBLIOGRAPHY

Chairman, Joint Chiefs of Staff, Manual, *Theater Engagement Planning*, CJCSM 3113.01A, April 2000.

Chairman, Joint Chiefs of Staff, Instruction, *Joint Strategic Planning System*, CJCSI 3100.01A, September 1999.

Chairman, Joint Chiefs of Staff, Instruction, *Chairman, Joint Chiefs of Staff, Commander in Chiefs of the Combatant Commands, and Joint Staff Participation in the Planning, Programming, and Budgeting System*, CJCSI 8501.01, April 1999.

Chairman, Joint Chiefs of Staff, Instruction, *Responsibilities for the Management and Review of Theater Engagement Plans*, CJCSI 3113.01, April 1998.

Chairman, Joint Chiefs of Staff, Manual, *Instructional Joint Strategic Capabilities Plan FY 1998* (U), CJCSM 3110.01A, December 1998.

Cohen, William S., Secretary of Defense, *Annual Report to the President and Congress*, 1999.

Cohen, William S., Secretary of Defense, *Report of the Quadrennial Defense Review*, 1997.

Lewis, Leslie, James A. Coggin, and C. Robert Roll, *The United States Special Operations Command Resource Management Process*, Santa Monica: RAND, MR-445-A/SOCOM, 1994.

Lewis, Leslie, and C. Robert Roll, *Strategy-to-Tasks: A Methodology for Resource Allocation and Management*, Santa Monica: RAND, P-7839, 1993.

National Defense Authorization Act for Fiscal Year 2000, Sec. 901, § 118, Quadrennial Defense Review, October 5, 1999.

Naval War College, *Resource Allocation, Volume 1, The Formal Process*, 1996.

106th Congress, *Title 10, United States Code, Armed Forces*, Washington, D.C.: USGPO, February 1999.

Schrader, John Y., Leslie Lewis, William Schwabe, C. Robert Roll, and Ralph Suarez, *USFK Strategy-to-Task Resource Management*, Santa Monica: RAND, MR-654-USFK, 1996.

U.S. Department of Defense, *Defense Planning Guidance, Fiscal Years 2002–2007* (U), April 2000.

U.S. Department of Defense, *Contingency Planning Guidance, Fiscal Years 1999–2003* (U), January 1998.

THE UNIVERSITY LIBRARY · UNIVERSITY OF CALIFORNIA, SAN DIEGO